ISBN (Paperback) 978-1-955364-42-3
ISBN (Hardback) 978-1-955364-43-0
Vets Publish
www.vetspublish.com

A is for Astronomy

The study of the sky

B is for Brightness

When the Sun is up high

C is for Crescent

A shape in the night

D is for Darkness

A curious sight

F is for Fascination

Eyes opened wide

G is for Glowing

The Sun's shining light

H is for Horizon

Where day turns to night

I is for Intrigue

As shadows appear

J is for Joy

Watching with cheer

K is for Knowledge

Learning so much

L is for Lunar

With a magicial touch

M is for Moon

Blocking the Sun's glow

O is for Orbit

A celestial dance

P is for people

In awe and trance

Q is for Questions

The curious ask

R is for Radiance

A celestial task

S is for Solar

The eclipse we adore

T is for Totality

Where darkness is more

U is for Umbra

Where shadow's play

V is for Viewers

Excited to stay

In every child's eyes

X is for Xray Light

A cosmic surprise

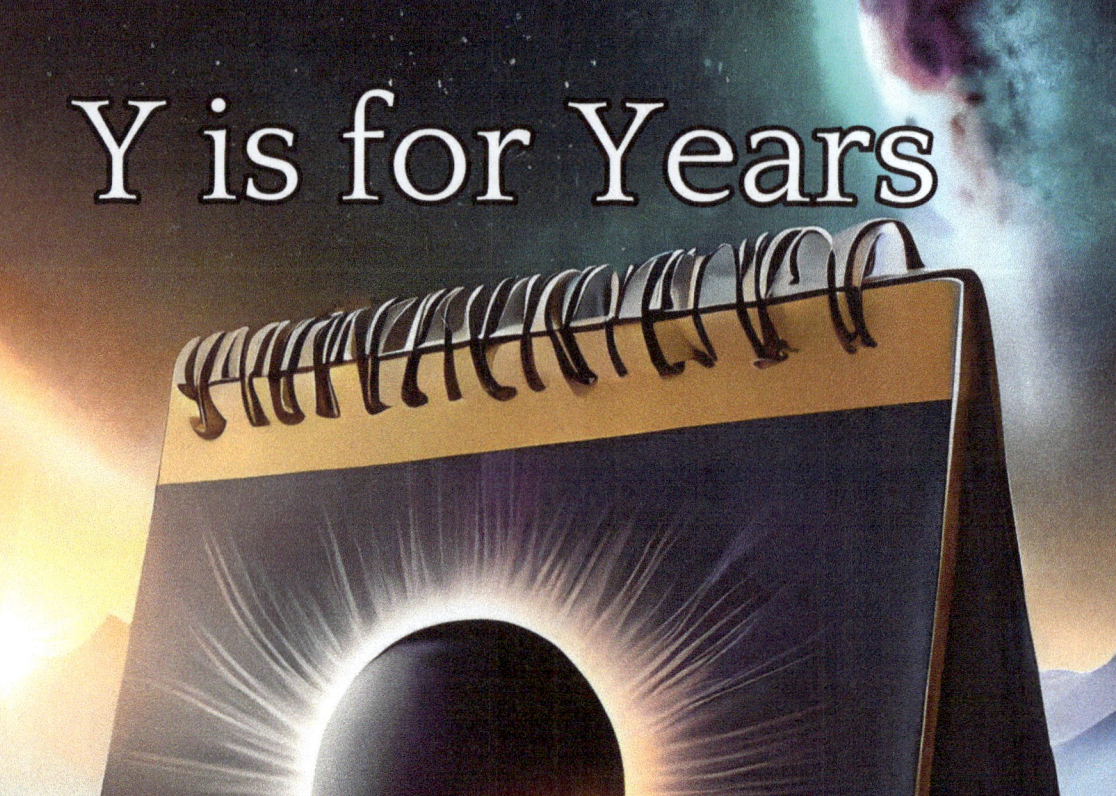

Y is for Years

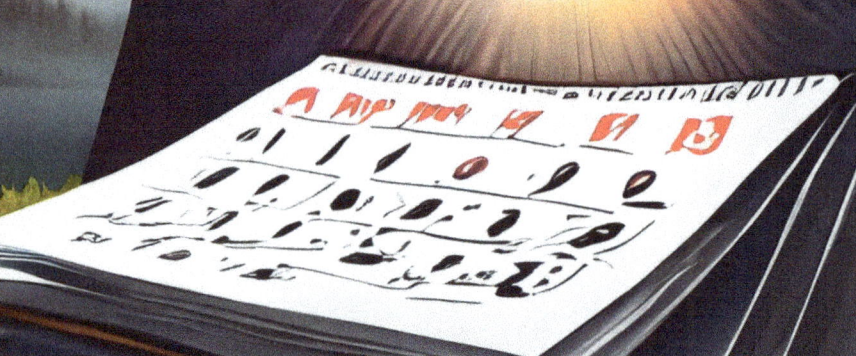

Until the next view

Z is for Zenith

Where dreams come true